This is the source, or beginning, of the Mississippi River.

Lake Itasca, Minnesota

The Mississippi River flows from north to south. It borders ten states.

Most of the Upper Mississippi River cuts through deep woods and rocky hills.

People enjoy visiting the upper river. What is this family doing?

The river helps make the land good for farming.

Barges loaded with grains and other goods travel up and down the river.

The Upper Mississippi is home to many plants and animals.

white-tailed deer

cardinal flower

mussel

Birds travel, or migrate,
along the river.

great egrets
great blue herons

St. Louis, Missouri, is a large city along the banks of the Middle Mississippi.

Many rivers flow into the Mississippi. Here, the Ohio River meets the Middle Mississippi.

Locks and dams keep the water deep, so that big boats can travel on the river.

Along the Lower Mississippi, the weather is hotter. Many plants and animals live here.

water snake

magnolia

alligator

nine-banded armadillo

Cotton grows
along the Lower
Mississippi.

Factories along the river's
banks turn sugarcane into sugar.

People ride in paddle-wheel boats on the river.

New Orleans, Louisiana

Near its end, the Mississippi
branches out. The land near
the river is called the delta.

Gulf of Mexico

This shrimp boat sails out into the Gulf of Mexico, where the great Mississippi River reaches its end.